Levis Reading Prize Entry
1997-98

Between Ice Ages

Bernard Gershenson

The Heyeck Press

© 1997 Bernard Gershenson
Library of Congress number 96-077642
ISBN 0-940592-29-0

Permission to use excerpts from Will Durant's *The Story of Civilization* was kindly granted by Simon and Schuster.

The Heyeck Press
25 Patrol Court, Woodside, CA 94062

for

Harriet and Charles

and for

Paula, Tasha, and Kate

NOTE

These poems don't have titles; they have epigraphs, all of which come from the first two volumes of Will Durant's *The Story of Civilization*. History is connected to memory and to our lives in a way that is more intimate than we usually acknowledge.

"Perhaps an earthquake shook the palaces into ruins, or some angry revolution avenged in a year of terror the accumulated oppression of centuries."

We've got a pretty good list
of hard times at home, not
the usual stuff for a country
free of war, but a few years ago
an earthquake scared the World
Series right off tv, and you escaped
the doorway that was supposed to shelter us,
running out the door past the leaning pine
that could span the road if it fell,
but it didn't, so I found you safe
enough for us to fight about what to do next,
if the earth shook again

And a year after that your school ran out of money
threatening to close, leaving you to teach kids
expecting a vacation at any minute, every night
a new meeting of angry parents, indignant unions,
callous gray men from the state capitol,
never before had we heard so much explaining,
all of us looking for the obscure detail
that would save your job, this was bureaucracy
as rocket science, detached, impossibly logical,
and not pleased at all, they said,
to be sending you a
layoff notice, but they did their duty, and we

waited for them to take you back the week
before school would start again, which they
without apology did. We understood this was
and wasn't simple luck and like
the earthquake wasn't personal. All part of
Life's Rich Pageant.

And after that the fire big as mountains,
the fire that would be history
shoved us into our cars with our
clothes and art and work and all the photos
that tell us who we were, the pictures that
may someday be clues to grandchildren or
great grandchildren or no one at all; and
five days later we came home, our house,
like your job, like our lives, spared,
luckier than neighbors a block away,
their houses vaporized, the unearthly smell
lying for weeks like a lazy cat
over these hills where we've put our lives

Of course, everywhere now there's blood turned sour,
a disaffected teenager with a gun,
a drunk starting his engine,
and you're out shopping at night
midst this relentless terror that
we must ignore, for this ignorance
is daily in our world,
everyone says so, psychologists
come from everywhere, the seagulls
after the locusts, to help us get on
with our lives; many people seem to think this
important

"When their civilization was already old—about 2300 B.C.—the poets and scholars of Sumeria tried to reconstruct its ancient history. The poets wrote legends of a creation, a primitive Paradise and a terrible flood that engulfed and destroyed it because of the sin of an ancient king. This flood passed down into Babylonian and Hebrew tradition, and became part of the Christian creed."

Imagine that! *In medias res* after all. We'll suppose
a big bang, popular and maybe true; that
settles how though we can't imagine before that
a faithless
belief in beginnings, conception and at some point
life, maybe then or not but in any case new, is it not?

It is not there are no new plots, for Sumerians, too,
could conceive infinity and like us not know
what to do with it, so strong a need for not only beginnings

See,
in our dream of nuclear holocaust
which we alone survive, except for someone
of the opposite sex, of course, with whom a phoenix
would be conceived with all the ensuing problems
of a wife for Cain, possibly
the descendent of someone else's dream

It's a bit muddled but all there somewhere
like the Sumerians wondering whether the Euphrates
would continue to rise over Ur

"The sunset at Gizah is greater than the Pyramids."

In midwinter the cold light through the window
somehow blossoms and makes cats of us all
shade becomes color a rosy
glow of our closed eyes over which
our hand is cupped in deference to
this everyday miracle which is no miracle
but a simple nap in the afternoon whose
dream pushes us over so that by accident
which is no accident we touch and set and
no one will notice how beautiful we are

"Since sacred tradition required that every Egyptian ruler should be son of the great god Amon, Hatshepsut arranged to be made at once male and divine. . . . To satisfy the prejudices of her people, and perhaps the secret desire of her heart, the great Queen had herself represented on the monuments as a bearded and breastless warrior; and though the inscriptions referred to her with the feminine pronoun, they did not hesitate to speak of her as 'Son of the Sun' and 'Lord of the Two Lands.' When she appeared in public she dressed in male garb, and wore a beard."

What can I say? The first club apparently the penis,
the first knife, the first gun, the first wand;
the cotton harvest comes much later and how incidental
though life itself; Hatshepsut looks in the glass
raking her smooth skin red with boar's bristles
herbs to please Amon swallowed, donned,
the miracle to prove beyond Karnak that
will never happen in two decades of new
markets, new pleasures for eye and mind
always a glance at the growing boy who wears
their dead father's name and will easily
bear his spear and sail Egypt everywhere it matters

Since it's your empire, no one's going to ask you
to explain yourself, but later on you ought
to know about an emperor who wore no clothes
on purpose, unlike the others

The governors are waiting and again wanting
slaves from Assyria or Libya for their stones
and the plans they have to please the Son of
the Sun which they have said which they have said
which they have said which they have said
and after all these years believe as does the Son
herself

"The horse is first mentioned in Babylonian records about 2100 B.C., as the 'ass from the East'. . . ."

She has made her hair the color of horse
the adolescent Arabian standing asleep
where it lives above the freeway
for her apple that had been dying
in the refrigerator near another freeway,
they've ridden the corral how long
certainly past suppertime, past parking lots
where sea gulls on vacation watch
from a fencepost, offering occasional
ill-tempered advice to which the girl responds:
crotchety gull, crotchety gull
isn't your belly ever full

the barn cat is mewling for attention,
for the simple world-weary open can
full of gunk that was once a fish, or many fish
another miracle—all the animals fit in the photograph
though it's a bit dark and fuzzy; we can't be sure
which girl it is or which gulls or which cat,
but the horse, he came in a truck on
still more freeways from the East
ages ago to find the others already angry,
obviously gods

"Hippocrates [the 'first' physician] tells us that '[Scythian] women, so long as they are virgins, ride, shoot, throw the javelin while mounted, and fight with their enemies. They do not lay aside their virginity until they have killed three of their enemies.' "

Some things never change: the myths of
our myths assume war equals sex and the two
cause each other; ask James Thurber or
Woody Allen or the Arab who invented
algebra; love won't be invented
for years, well after husbandry and France.
If you read enough footnotes, you'll learn
everything you didn't know you didn't know;
apparently civilized men have been terrified
a long time, for who else would the enemy be
with no reason to ask what do women want.
Democritus, who liked to travel, invented
atomic theory and maybe hearing this story
brought back tales that could be true, I mean
Hippocrates was only a doctor; what did he know?
Now, even as we speak, some sex or other
has lived to overcome Scythians they've never
heard of and who wrote nothing down
but had to exist for us to understand
what we don't understand; is it any wonder
we have trouble believing the most
respected physicians?

"The Second Commandment elevated the national conception of God at the expense of art: no graven images were ever to be made of him."

"It was the usual thing for ancient law-codes to be of divine origin."

A cross between the head of
an owl and Michelangelo's
Moses which I'd seen
a picture of somewhere
great white wings over
the whole earth but his head
could only be in one place at a time
now there's a problem
or at least an explanation, for
only Mom had eyes in the back of her head
she said though I figured they couldn't
see through the hair and anyway,
like Dad, he was busy, he had
meetings
so that's why fair's fair or
life's not fair depending
on just do as I say, and I don't
mean later, "Oh God,"
"Don't say God, and don't
look at me like that. I can
see you. I have eyes
in the back of my head."

"A nation, like an individual, begins with poetry, and ends with prose."

While I love you frogs
  are disappearing mysteriously
    from nations everywhere in

the world we were given
  as caretakers for children
    we can't assume yet imagine

but frogs we don't think about as
  empty spaces in anonymous summer
    with cow brown skies that

settle on rags in grocery carts
  pushed through a reef of mumbling
    litter by spittle and cough

waiting too long for the princess kiss
  murky ponds and tasteless mosquitoes
    but no one thought them hopeless

Night hides the red eyes we didn't look for and didn't see.
And frogs really are disappearing. It's been on the news.

"Truth does not often escape from palaces."

". . . Bindusara [ca. 290 B.C.] was apparently a man of some intellectual inclination. He is said to have asked Antiochus, King of Syria, to make him a present of a Greek philosopher; for a real Greek philosopher, wrote Bindusara, he would pay a high price. The proposal could not be complied with, since Antiochus found no philosophers for sale."

The oily voice oozing from the radio
tells us to fear everything, especially
taxes, especially each other because
they're all out to get us; trusting even
the image in the mirror might be risking
the safety of the locked bathroom; to prevent
the sky from falling, he advises,
we should buy many guns; then he sells
us steaks from some mail order place
in the Midwest. His audience
praises him.

The mailman delivers
what we never asked for, appeals
and promises, coupons that have come
all the way from Minnesota, explanations
of money, insurance against loss, recipes:
money in sufficient quantity can dress
any wound, you see, we all believe in it;

that there are those who would be President
or King; here, in the kitchen
we think we rule our children
and they're enough.

"Mumtaz gave her tireless husband fourteen children in eighteen years, and died, at the age of thirty-nine, in bringing forth the last. Shah Jehan built the immaculate Taj Mahal as a monument to her memory and her fertility, and relapsed into scandalous licentiousness."

more of a bluish blue, indigo, smalt
the way of her body, watchet, zaffer
the next one deep in the earth, cobalt
and every year more bloom more blue, sapphire
the turquoise young, the sky reflecting more
bellflowers, bluebells, even penstemon
I was twenty, my vision azure for
my only future, for me blue lupine
everywhere I loved the word cerulean
my days were children my ocean my days
and my sheets were my days my thigh's vein
he knew he would kill me, his body a haze
before I was thirty-nine I was earth
facing blue, cold as tile, lost in birth

"The word *chess* comes from the Persian *shah*, king; and *checkmate* is originally *shah-mat*—'king-dead.' "

That would be the end except another
game later or tomorrow or by postcard
we leave his majesty to rue predicament,
the live dead king; maybe in exile
waiting for mail, memoirs of oversight
the unexpected queen all her life
thought a pawn; he couldn't
have known but should have seen, or
maybe he's dead on the shelf in a closet
waiting for you the grand vizier after
dinner with an aristocratic guest who also
knows knights are hard to control, the church
is never straight and the queen doesn't need
to bear children
yet
does he die of humiliation or warts,
or do elephants pound him into tender
memories when he sings his little song

"The unresponsiveness of a god is no obstacle to his popularity."

The chiropractor says reader's posture,
my back might kill me but I've a fine
mind, the girl cuts her finger instead of
cardboard, the deepest red ready for
puberty, her mother's stomach hurts
again, older sister asleep in the shower
bruises yellowing her knees, the brother
never got conceived so that as they say
was that; some are picky eaters
the mother says don't let me
drink coffee at night but who
would dare try her, there's plenty
of waiting, of course, a refrigerator
that Darwin could have studied
did anything interesting happen in school today
who wants to know, aren't some things
sacred. Kate says do you believe Elizabeth
loves God more than her mother?

"It is almost a law of history that the same wealth that generates a civilization announces its decay."

The spirit or the letter of almost? Like cancer and
auto accidents, someone else and besides
only from some university, prophets
aren't new, we don't even think
of ourselves as handicapped, so then, light years
from here practically speaking, videotape of the
overturned truck with the cargo of polyurethane
popcorn, a plastic blizzard in the rush hour wind
that subsides to flurries and disappears like
discarded Coke bottles that have learned to bounce

". . . men will pay any price for mythology."

in time, the holy spirit was seen in the internal combustion engine, and many religious denominations competed to see whose spirit was most powerful and which could build the most elegant temple for the spirit to drive. Each man and woman coming of age sought the companionship and protection of his or her own finely tooled spirit, and some families in an attempt to achieve oneness with the creations that feasted on fuels made from the stuff of the earth's first creatures collected as many as possible. And those who could not afford their own spirit were looked upon as deprived, and those who chose not to give themselves up to the spirit, like many who lived in Manhattan, were looked upon as atheists or communists or both or worse and were not trusted by the believers who were fruitful and drove their spirits throughout the land where there were spare parts.

"The goal of philosophy is to find that secret, and to lose the seeker in the secret found."

the bird clues:
the rufous-sided towhee sees the white cat
the Oregon junco won't eat seeds from Safeway
the Stellar's jay misplaces mountains
the red-tailed hawk, hawk, hawk
the albatross spills salt

the toxic waste clues:
egg-shell paint in the chicken feed
petroleum thinks it's your brother
everyone knows polyurethane
water tastes like the army

passive clues:
fruit was eaten
the box was opened
ships were launched
women were blamed

". . . Portugal sent a fleet to India, with instructions to spread Christianity and wage war. In the seventeenth century the Dutch arrived, and drove out the Portuguese; in the eighteenth century the French and English came, and drove out the Dutch. Savage ordeals of battle decided which of them should civilize and tax the Hindus."

after thirty, meaning comes to property
no one pulls out a wallet, puts a hundred
grand on the table, asks for a deed please
and a receipt thank you, it's a sordid
business buying between the numbers;
counteroffers appear like dragonflies
but here, good help, even the bumblers
cheating with insiders, spiders spinning
for the nature photographer narrated
by a sort of famous guy, clearly,
these folks take money seriously
and they keep on taking it
up the elevator to a title
company made out of the stuff:
the administrative assistant
gets to say sign here, please, and here
and here and here, and who's to
argue with the Leibniz of legal pads;
through her we enter file drawers
we hadn't dreamed existed, the simple

deeds from monopoly, each with its color
each zoned for not only houses but hotels,
the word easement not mentioned once,
escrow a mere trip to the bathroom,
such easy rules, as easy as conquering
India

"P'an Ku, the first man (they tell us), after laboring on the task for eighteen thousand years, hammered the universe into shape about 2,229,000 B.C. As he worked his breath became the wind and the clouds, his voice became the thunder, his veins the rivers, his flesh the earth, his hair the grass and trees, his bones the metals, his sweat the rain; and the insects that clung to his body became the human race. We have no evidence to disprove this ingenious cosmology."

You know yr almost up to geology,
almost down a few feet from the canyon's rim,
and that gnat, that louse, that mosquito,
well, friend, they're you, frog food we didn't anticipate
but, hey, not yr fault, that first rhythm a scratch of the
head that took a few years, look from a hill on yr block into
sky, unsullied blue, it's not far enough, from space then,
the free floating astronaut's peaceful planet, still not
far enough, the smooth, pink womb conceiving the Nile, the
Amazon, the Yangtze, no wonder yr so impatient, all that space
and no time to do anything but cause a rash, raise a welt or two
ride out to sea in a salty teardrop, still not far enough

“‘When I have presented one corner of a subject to any one, and he cannot from it learn the other three, I do not repeat my lesson.’ Confucius, *Analects*”

Well, then, how about a hint? Just kidding like the judgment we all want, enough
kidding to recognize loss, of a watch, a
tree, a love; only is fair, my love
for you make me feel better. I am
purposely obscure, you see, giving
only the one corner, every thing nameable
drifts into metaphor even though I
see, the rabbit yes from the hat, the
red gold of your hair at four o'clock on
a sunny winter's day in these hills.

But in
bed at night, while we absorb the day's
small disasters, the bad plumbing more important
than someone's war, the kid's backtalk more
insulting than politics, the dirty dishes more
ominous than a bloody cough, lines we
won't see for months etch themselves
around our eyes and mouths; we wrestle sleep,
Jacob's angel, our bodies, exhausted, ignore
each other, and the angel throws us into the
one corner we know where
we dream of the other three.

"The Huns, barred for a time from Chinese soil, moved west into Europe and down into Italy; Rome fell because China built a wall."

Because Shih Huang-ti built a wall
Neil Armstrong on the moon
knew earth the only thing out
there, the thing
with a wall a wall to keep
to keep out the bad guys the bad guys
who turned who turned and rode the
other way a long way that
Armstrong saw the whole way
all the way he saw on the earth the earth
that began at a wall that was China he
knew he knew from the moon Shih built
he built so Armstrong could see could
see the wall of dead men and women
dead as the moon of weightless light
and weightless dark the dark no one
saw no one no one saw

"... Ming Huang ... ruled China, with certain intermissions, for some forty years (713-56 A.D.). He was a man full of human contradictions: he wrote poetry and made war on distant lands, exacting tribute from Turkey, Persia and Samarkand; he abolished capital punishment and reformed the administration of prisons and courts; he levied taxes mercilessly, suffered poets, artists and scholars gladly, and established a college of music in his 'Pear Tree Garden.'"

Some day, all this, the polished bamboo flutes
Tigers loose in rice, news from caravans;
An opera of elegant trumpets and lutes
On horse or ass or camel, Uzbekistan,
Ankara, Persia, new wealth of the West,
The bandits we catch, the armies we feed
To wish and law, our civil service test,
Our opera, garden, pears, actors, jade;
But these are notes in pentatonic scale
Of the peasant we've eaten for breakfast
With ancient teas: jasmine, ginseng, camomile,
Even rice, whatever will please a guest
Or a concubine or the long green snake
Whose young will inherit enemies we make

"An Lu-shan's barbaric hordes sacked Ch'ang-an, and slaughtered the population indiscriminately. Thirty-six million people are said to have lost their lives in the rebellion. In the end it failed; An Lu-shan was killed by his son, who was killed by a general, who was killed by his son. By the year 762 A.D. the turmoil had worn itself out, and Ming Huang returned, heart-broken, to his ruined capital. There, a few months later, he died. In this framework of romance and tragedy the poetry of China flourished as never before."

If the blood of each were a unique color; if the girl
riding her bicycle to school rode violet, say, the man
selling hardware running silver; if we could change
the color of sky, of grass, flowers, mountains, the
works, these would be issues of war, and millions
would lie open till one was found whose life
was the color of sun, and another whose
veins flowed night, and every artist's canvas
would show a separate peace for all those
bloods whose colors would be seen in sentences

If each color were someone's blood; if the pictures
we looked at were the art of lives we could look for;
if every shade were a new metaphor and each
bloody wound a signature only the author
could duplicate; if we loved sight like sex

" 'The state,' [Emperor Wang An-shih (1021-86)] said, 'should take the entire management of commerce, industry and agriculture into its own hands, with a view to succoring the working classes and preventing them from being ground into the dust by the rich.' "

"Though most of the taxes were taken from the income of the rich, part of the heavy revenue needed for the enlarged expenses of the state was secured by appropriating a portion of the produce of every field. Soon the poor joined with the rich in complaining that taxes were too high; men are always readier to extend governmental functions than to pay for them."

The teen-age American army wades through
rice paddies and rain forests, just like they
did back home in Detroit, McCook, Concord,
McComb, Gallup, Salt Lake City, so they
understand everything; the buffalo people
with no Marx brothers or Mayflower
Van Lines around the shell that housed
Lenin, always waiting in Finland. Do we
understand this, a fight about who
gets the rice in the Mekong Delta, the poppies
of old Siam? Wang An-shih reads Tu Fu, Li Po, Po Chii-i,
sits down, invents morality, as he or anyone
should, and by coincidence nine hundred years later
I come up with the same ideas; if only he'd taken
better care of himself. Thirty years ago a bunch
of us were in high school; and three years later,

dead in a land we'd never heard of. We'd heard of
potatoes; ok, potatoes and
spaghetti. It's still that way when the sun
sets, supper appears on the table, and the blue light
of television tips us off about the end of the world,
the headache remedies we should buy, the reasons
Wang An-shih lost his job

"About 105 A.D. one Ts'ai Lun informed the Emperor that he had invented a cheaper and lighter writing material, made of tree bark, hemp, rags, and fish-nets. Ts'ai was given a high title and office by the Emperor, was involved in an intrigue with the Empress, was detected, 'went home, took a bath, combed his hair, put on his best robes, and drank poison.'"

On paper your simple lines may betray
A guarded heart defending secret songs,
We only sang the night we sang till day.

We tied our words in simple knots to say
That unsaid prayer that somehow belongs
On paper; your simple lines may betray

Us if cormorants drop their fish to lay
Their catch beneath our star, beneath our tongues.
We only sang the night we sang till day

When last you signed the nets that found their way
To court where jealous birds sought easy wrongs
On paper your simple lines did betray.

If this is love we've written just to slay
The Son of Heaven, feel death in our lungs,
We'll only sing tonight; we'll sing till day.

Hemp, bark, nets, rags, obviously astray
With words of disgrace, the thoughts of Ts'ai Lun
On paper; your simple lines did betray
Our only song the night we sang till day.

"When two European painters accepted the invitation of the Emperor K'ang-hsi to come and paint decorations for his palaces, their work was rejected because they had made the farther columns in their pictures shorter than the nearer ones; nothing could be more false and artificial, argued the Chinese, than to represent distances where obviously there were none."

From this window I can see
thirteen
pine trees,
one
spruce and one
redwood which
is younger than I am, skinnier
but
much taller like
Abraham Lincoln
whom I cannot see
in California where
he never was
tall or short in years
that finished their
yearness and he
was dead and
removed to pictures
painted, some of which
did not show
all of him;
all of the trees
in the window

are the same size from
our bed at night
and above them
the stars that
have been the same
size all our lives
though the ones
in the window
belong to us
as do the thirteen pines,
the spruce and the young
redwood. The kids,
in another room,
talk constantly to
friends they can't
see but believe
are talking to them. And
Lincoln is memory
of memory, someone all of us
within borders we believe
but can't see
believe did something
that all of us should know
possibly just the beard
in his picture in
a book that is
usually closed on
the shelf that
collects dust we
can't see at night

"On [Marco Polo's] deathbed his friends pleaded with him, for the salvation of his soul, to retract the obviously dishonest statements that he had made in his book; but he answered, stoutly: 'I have not told half of what I saw.' "

Me neither. You'll have to wait.
My story has been heard before,
but maybe not by you, or perhaps
you won't recognize it in its currency,
the years crossing barren landscapes
populated by jackals, sheep and bandits,
the shadeless deserts, my beautiful wife,
not yet my wife, tied somehow to railroad
tracks for God knows what reason, her kids
thankful they knew where she was though
a bit unsettled by her preoccupation, their
lives, understand, crystals of curiosity,
patience and denial, cats, turtles and moths,
the wild narcissus that apparently
grow everywhere, thorns in the paw,
the crown, but I see you don't
believe me. Shall I tell you
about the Onanist from the Dominican
monastery at the end of a New Mexican
road where snow fell all night as he
cursed his father in his sleep? Or
about the young mother escaping

her husband and deserting her kids
who cried while she slipped her hands
over me in the back of a bus? She had
a friend in Columbus, she said. Or
about Mr. Jones, my sixth grade teacher,
who believed the civilized world depended
on long division. The Russians, of course,
had placed a moon in outer space
and everything surrounded it. If you
make me recant these stories, if you
need them fiction, I will cross my fingers
behind my back.

"When European capital built the first Chinese railway (1876)—a ten-mile line between Shanghai and Woosung—the people protested that it would disturb and offend the spirit of the earth; and the opposition grew so vigorous that the government bought the railroad and heaved its rolling stock into the sea."

Every day I walk the mail where it wants to go
sometimes with a lute or a trumpet,
sometimes in silence, the earth a drum under
foot whose music warms the globe,
the heat at times unbearable, or worse,
barely felt through a muffle of snow;
the thoughts I carry settle into their
imperfection, apprehensive love,
notices to appear before the magistrate,
remembrances of a father lost in
a foreign city, thoughts dreaming
of their reception, asleep with
the rhythm of steps beating our
silent, breathing drum.

Conversation
would pass me by: ". . . before the
cow kicked uncle . . ." "If she has
another girl . . ." ". . . has failed
his examination; . . ." ". . . the
Emperor doesn't know we . . ."
I could hear goats walking my way.
And then

came the steel net on the earth,
the metal harder than taxes, and
on it came a locomotive, whistling,
snorting smoke, rolling in its
flatulence, hissing louder and faster
than thought. It drank more water
than summer, ate more coal than
winter. This was not the dragon
of the new year or a snake to eat rats;
the dead were confused; love
lost to noise, everyone could see,
so we petitioned the Emperor
to drown this foreign monster
before it wakened every soul
that had told its story. I carried
the letter myself.

"It was the custom to rank scholars, teachers and officials as the highest class, farmers as the next, artisans as the third, merchants as the lowest; for, said China, these last merely made profits by exchanging the fruits of other men's toil."

Of course, laughter comes to mind
I one Saturday cracked a rib
tripping after a frisbee.
The air was sour mash
the distillery near the smokestack park
a small grove of candy wrappers and empty
aluminum crumpled in a strange and small
pleasure instantly forgotten, two boys
leaning against wire around an antebellum
oak, a brown bag passed from nose to nose
falling uninjured, laughing, happy-go-lucky
students on their weekend by chance
meet a teacher with his ridiculous injury
all of them caught in a bubble
incidentally fermented. This is near
where the young Cassius Clay learned
to box some years ago now, the smaller boy
forced a giggle and said we ain't afraid of
you; aren't, I said, we aren't afraid;
of course, now, their laughter comes to mind

". . . those who succeeded received the degree of *Hsiu ts'ai*, entitling them to membership in the literary class. . ."

Down the street a black and tan hound
follows a man balancing a trunk on his head
and in the trunk are the secrets
of his life, the stuff of his future
or a past he'll discard by the side of the road;
either might be dirty linen or electrical
marvels which language does not yet accommodate;
the hound's tail wags, squirrels
scatter and imperiously scold but neither
dog nor man appears to notice
witnesses, yet they will
be graded, judged, audited, and if
found worthy, if body and mind are
still intact, if there is room

Later, as the sun softens the asphalt
men from the water works stop their
truck and perform a rite only they can know
under the street where there are
no dogs or humans carrying trunks;
the sparse traffic must circle the
empty truck and the barrier with the warning
men and equipment working;
nowhere is there a sign *keep*
out, it isn't necessary, for
we know our entitlements, when
we've earned the privilege of looking
at the dark side of the street

"Typical tests were a composition of a poem on the theme: 'The sound of the oars, and the green of the hills and the water,' and the writing of an essay on [a] passage from the Confucian Classics. . . There was not a word in any of the tests about science, business, or industry; the object was to reveal not knowledge but judgment and character. Those who survived the tests were at last eligible for the higher offices in the state."

In September, school opens, the harvest
is in
the yellow bus on the flat county road
scattering crows
pecking at something on the cool blacktop,
they can't
see town but they've been,
I think,
though they don't stay long,
just the time it takes to hear
someone else's decisions,
you recognize at least part,
of course, the art counter

at Woolworth's where the kid
who'd be embarrassed to shop there
rings up the thumb tacks, a colander
and school supplies; no one has ever
seen anyone stop
at the art counter.

But you've seen the paintings
in the disinfected motels
that smell like nothing else,
on the walls of the local knitting shop,
in restaurants whose breakfast menu
includes hot biscuits and gravy,

they're purchased after
hours

this year there are two new kids
to ignore unless one has brought a lunch
worth trading for, wherever
they came from, they'll say,
is better than here

the main idea of the paragraph was:

a. Billy was afraid of the state of Illinois
b. Billy was taught to distrust Ethel
c. Billy Billy Billy
d. All of the above

"Takatoki . . . had a passion for dogs; he accepted them in lieu of taxes, and collected from four to five thousand of them."

Somewhere in the foreground
the streets of the neighborhood
Sherwood, Merriewood, Robin Hood
Way, Nottingham Way, piney woods ready to fall
on you and the Merriewood Doggie Chorus
that sings at night for this the only
house defended solely by big trees
that never fall on the mailman bringing
bills in his jeep and the news
that Mother
is on her way; just yesterday
a large black dog bit
the right front tire

Randy the mailman knows return addresses
like you know your way home; he's solved
every mystery on his route of which I am one;
Randy knows your friends and creditors, your
employer, your politics

He's appropriately glum giving
you an overdue notice, and he'll
ask about the theater you subscribe to

Sandy the cocker spaniel sleeps in the road
on sunny days; Sydney the Australian shepherd
misses Bubba and Freckles, whose time has passed

Daisy barks when her people leave, her throat
perpetually sore, and Cody, the Rottweiler next door
comments to everyone every time
I pull a weed

“Literature flourished and morals decayed.”

collecting dust
above the desk on a shelf
three beanbags, vaguely
reminiscent of penguins, sit next to
Alyosha, Ivan and Dmitri
the shelf a strange bed
yet our bedfellows are comfortable enough
trees, cotton, legumes and whatever
they make ink with, Alyosha, still
innocent, sees innate goodness
in the beans, the shelf, the unread
nine hundred and forty pages, some of which
contain his name, but Ivan, the intellectual, imagines
a higher purpose; he’d like to see all
the beanbags in the air at once,
but he won’t be responsible for
where they land and Dmitri, the hedonist,
won’t care; trying to see
what he will never see, he pulls himself,
his brothers, the whole of the Russia he knows
onto the beanbags. Everything (dear reader)
slips to the floor and waits
for order

"We do not know which of the many roads to decay Crete chose; perhaps she took them all."

Drinking our dark rain that burns trees
And fish but surprises no one
We become map, a picture of choice, neurons we
As impulse run over breathing invisible
Brown stuff that makes us a strange red rash
On the arm, itching, which map calls legend,
Fat red lines faster
Than skinny blue lines which run to
More major organs than skinny gray lines
Whose access is uncontrolled, free, any-
Where love goes frogs, locusts, boils

Forgive me. I have lost track of myself
But always I've looked out the window
Depending on a benign sky for help,
Interference from the saddened cosmos
Lost love, spilled oil, killings of all kinds.
You curl tired into your chrysalis
And shake my touch into the dying pines
Also outside the window; we are Sisyphus
Half asleep, sighing through relentless news
Of my desires, the epidemic a-
Round us, the student who doesn't like you,
Discord and death in Stockton and China;
Your tense and tired muscles twist the sheets
While a siren pushes night through the streets.

"As if to please the modern spirit, more interested in plumbing than in poetry, the builders of Cnossus install in the palace a system of drainage superior to anything else of its kind in antiquity."

In the parking lot someone has placed
a flyer under your windshield wiper:
'Poems Supplied and Repaired—Haiku to Epics, Free Verse or
the Form of Your Choice—Critiques—No Subjects Are
Too Insignificant—Specializing in Love and Angst—
Emergency 24-hour Service—
Reasonable Prices—Will Deliver'
There's a phone number and a logo:
a quill is lying across
a computer keyboard at an
obtuse angle

When you get home, the same flyer
has arrived in your mail box;
at last there's somewhere to turn
when the poem just won't flow,
when the muse is on maternity leave,
when writer's block is Albania;
it turns out all lines are busy,
muttledom,
the music from the phone *Pomp*
and Circumstance
still life of shopping lists,
appointment books and other

important papers waiting for
the line to clear,
your turn,
your poem;
a man wearing a jumpsuit
arrives in a white van complete
with auger, ledger and motor;
he checks your journals, squirrels
through the wads in the wastebasket
which smells of sulfur,
says no problem, attaches the motor to
the auger, hands it to you and says
swallow; it tastes of some fruit you
can't identify; then he hands you a toggle
switch and says whenever you're ready;
you feel the buzz of Pandora's box
reaming through entrails that
turn out to be yours, and it's over,
this late twentieth century
with its modest advances. After
you pay your bill you try out
your stuff, and a perfect sestina
falls out of your brain the way it's
supposed to; later, on the phone,
your friend sympathizes: it's
such a pain in the neck when
all the pipes back up, and the cost,
you should have been a plumber
everyone says but no one means it

"When his new wife bore him children [Schliemann] reluctantly consented to baptize them, but solemnized the ceremony by laying a copy of the *Iliad* upon their heads and reading a hundred hexameters aloud."

For family outings, she said, my father
took us to the library; the amusement park
of labyrinthian stacks and more stacks
of musty adventures, yellowing pages
and the special dust that readers breathe,
the rooms without pictures where he would fade,
while I sat at the table in the wading
pool of picture books, Ping fishing with his
cormorant, Officer Clancy making way
for ducklings; I'd bring home as many as I could,
never enough; but we'd go back; how about
a ride, he'd say, and we'd ride to where books
were waiting like the Grand Canyon
till I could enter the room with the smell
of secrets, the living room of strange facts
which he sometimes revealed at supper while
buttering and buttering and buttering his bread.
And I would tell him what I had read.

"We are told that [Terpander] ended his life at a drinking party; while he was singing . . . one of his auditors threw a fig at him; which, entering his mouth and his windpipe, choked him to death in the very ecstasy of song."

Anticipating the next note, it might be a time
of winter rains, a dark afternoon, Spartans
on leave for a change, everything sweet
to cloying, the good times, a lamb
in every pot, pomegranates
to suck and spit, reminiscences
of summer fields
of ripe Helot women and lithesome youths ready
to wait for their next shaft
from somewhere, the older man who might conduct
a choir of swords and shields, the only singers
native born

But Terpander of Lesbos sings of the grape.
Through an oversight, the latest war has ended,
so some nameless hero takes to drink, and lo,
he waxes obnoxious; drink will do this, you see,
cutting off blood to the brain, the soul,
full of sentimental sediment, creates gods
that no one else hears, soldiers restlessly
count until the next rumor of rebellion relieves them
of songs they can't sing

In another part of town striking bus drivers
remember the scab who drove over their
friend who'd been innocuous since high
school. "Well, I'm just a bus driver," he'd say, a
lot of the bus drivers said that; he didn't have
a business card but did have a wife who
worried about drunk drivers and psycho-
paths riding to Visalia or Bakersfield,
men who wouldn't obey no smoking signs,
who snored into someone with sharp elbows

In this country in those places
juke boxes do all the singing,
music no one hears to drink beer
and eat pizza where some wise-ass
hits *Jambalaya* by Hank Williams six
times in a row, and all of a sudden
everyone knows "Come on Joe, we gotta go
Me-o my-o" is coming round again

"So [Heracleitus] went off, like a Chinese sage, to live in the mountains and brood over the one idea that would explain all things."

On days without death, nothing is profound:
the homeless gather quietly by the culvert
they'll wash in the creek that's low in the summer,
and they'll piss in it until the town
declares a health hazard, the idea to explain
all things, the inability to breathe all
the fresh air anyone could ask for. How
dare they show themselves

This one uncle would save his niece, dip
into his bag and sip the slow bottle why not
disappear into the sack where he'll eventually
bury himself—why not save himself this
way, afraid of dogs, but it isn't every day
you and I talk of these things or even think;
like ants with their amazing chemical drive,
we know where we're going, and no one
needs to explain anything
to us

> "Male posterity avenged itself upon [Sappho] by handing down or inventing the tale of how she died of unrequited love for a man."

My daughter, whose skin is incidentally white, keeps company with a boy whose skin is also incidental but not white. Their laughter, neither incidental nor any color at all, leads them like a guide dog into houses and rooms and each other's arms. It's hard not to be afraid for them since they don't know to be afraid for themselves. Should they part as all young lovers do, who will rewrite their story?

"... the Sybarites marched out to battle with an army of 300,000 men. The Crotoniates ... threw this force into confusion by playing the tunes to which the Sybarites had taught their horses to dance."

This history is repeated every morning the
world goes to work; the man in the rear-view mirror
shaves stubble into his briefcase; the woman in
the next lane does something to her hair with
an indistinguishable implement that might have
been discarded by someone's secret police;
the voice on the radio is singing a song of
traffic, weather, the partly cloudy twenty
percent chance harmonizes with sales of many
fine products and on this station at eight, eight
on the freeway, we hear the *Washington Post March*
just the way John Philip Sousa always imagined
it would be played.

No one admits to confusion. After all, we know
where we are and ostensibly where we're going.

We breathe all kinds of invisible crap,
the crawly slime from childhood that,
it turns out, is as middle-aged as the commute;
in the car over there, the driver lights a
cigarette, and suddenly I feel sorry
for us all.

For no reason I can think of I think of
you in your car on a route that only you
can take; I think of the battles you'll
tell me about at dinner; I doubt that
you hear Sousa trying to wake up this
sullen army of unsmiling workers.

I am possibly the only one listening to this
radio station but not the only one tuned,
I am sure of this and smug this way.

So maybe I
am the only one confused, maybe I
am the only one wondering where
the celebration is and whether I
am pointed in the right direction.

"Since all bodies moving in space produce sounds . . . then each planet in its orbit about the earth (argued Pythagoras) makes a sound proportioned to its rapidity of translation, which in turn rises with its distance from the earth; and these diverse notes constitute a harmony or 'music of the spheres,' which we never hear because we hear it all the time."

The music we don't hear must be deafening,
louder than war, louder than television,
louder than a ball through a window

The music we don't hear must not care
whether we dance, whether we call out,
whether we know the lyrics

The music we don't hear, a shadow that
falls in the forest, spills into night,
waits for the low light of winter

The music we don't hear is witness to
our version of God, our hoping for hope,
our listening for angels in the area

"Ares was never distinguished for intelligence or subtlety; his business was war, and even the charms of Aphrodite could not give him the thrill that came to him from lusty and natural killing."

In front of the mirror he stands naked
admiring the tip of his sword, the black
hole of his gun, the head of his penis;
this, he says, is symmetry

He looks at Aphrodite's cleavage;
sex is good, he says; it makes us forget.
But war, war makes us remember.
Now I ask you, which is better?

And sex has another problem, he
goes on—then repeats himself slowly:
Sex has another problem (if it's to be good)—
You need a willing partner!

Most men prefer women for this role
And any man who's honest
Knows the most malignant enemy
is less foreign than woman!

And he slams his stein on the table.
Barkeep, he's yelling now, Let's have
another round. Rape, rape is good, too,
but let's face it—that's jerking off a war fantasy.

Why mess around with rape
when we can go to war;
an enemy is easy to invent.
Unity depends on the enemy.

He smiles at his profundity. You think
me insensitive? Not so, not at all.
Emotion is everything, and what, I ask you,
Gives us more than killing and killing again

and risking death and watching compatriots die,
watching their blood, creeks, streams, rivers of life
run back to our mother the earth—
what agony, what anger—this death is life
I can watch for years at a time

When do tears flow more freely?
When is laughter more necessary?
When are fear and courage more visible?
When and when and when?

And imagine the gratification, the satisfaction,
Everyman a god, imagine
as you disembowel the enemy, the victory
that glorifies and hallows simple blood

If that's not sensitivity, I have once again
drunk too much and whet my tongue on grain
from your cracked and thirsty plain; I know
you didn't believe I understood these things.

"There was a pretty legend in Greece that the first cup was molded upon Helen's breast."

After it's not cancer, after the senseless
walk to the car from which baked air
wakes you into the mechanics of running
heavy machinery into midday

After it's not cancer, you forget
the speeches you've been rehearsing
and sit embarrassed for yourself
but it's easy to forgive yourself

After you're moving again the buttons
on the radio push the same stations
that play the same music and report
the same news that you ignored
an hour ago

As it crushes people you've never met
who haven't heard about you and your
fresh air, your perfect hair, your
long blue drink of water, who

haven't heard the world is fair

"Comedy, says Aristotle, developed 'out of those who led the phallic procession.' "

Where are the leaders of yesteryear
who knew how to tell us a joke, make
us laugh, or maybe when they say let's fight, maybe
they mean for us to pick up cream pies, seltzer bottles
Why shouldn't we laugh at their erections
Why do we believe they want to fuck us over;
Dangling between the legs, it's not pretty
bumping along on its balls, the sagging skin
of the elephant, and then, sometimes without
warning, triggered by godknowswhat, the nose
of the dog pointing at its prey, the hose thrashing
on its own, and we men will follow this brain-
less leadership anywhere, over mountains, through bad
weather, down to the kitchen floor, true slap-
stick, and the joke's always a good one,
we think

"The Greco-Persian War was the most momentous conflict
in European history, for it made Europe possible."

Divorce, we know, engenders bitterness
and weariness
Where were you while I waited in
the cornfields with my thumb out,
the cars as scarce as landscape
Would you have helped me hop on
one foot, arms stretched birdlike off
in all directions, that is, at least four, that
simple crossroad, from where Wisconsin
might as well have been Bogotá,
first on one foot, then rain came
and nothing, not even wasting time,
was simple anymore

Now you know why I missed the news,
the high tide of history passed me by like
prayers someone must have put in
for travelers; until then I had
thought myself observant—the rain, for
example, I recognized right away
as trouble

"We have no testimony to Aspasia's beauty, though ancient writers speak of her 'small, high-arched foot,' 'her silvery voice,' and her golden hair."

Each miracle here the work of consciousness
on a small plot of land where tomatoes grow
most of the year, though they're not always
good; in the tomatoes, in the plants, in the
soil surrounding, the periodic table
ebbs and flows like desire on a day
that becomes progressively hotter
until the feet swell, the tongue festers,
and the most we hear is a mumble,
but no one records days like these or
remembers whether any great ideas or good
ideas ripened and lightened our hearts.

It is nap time in our part of the world;
we throw a leg over our partner and dream
even more confusion than we've just seen;
lying in bed, awake, we're so aware we're
thinking we give up all hope of intelligence;
"Nothing will come of nothing," as Lear says.

Hair in this weather is dead grass; crickets
sing and cicadas; flies make a world of
irritation, and all thought focuses on
dry, sticky lips; even the good thoughts
aren't new enough to say out loud.

"The freedman is especially valued as an executive, for no one is more severe with slaves than the man who has come up from slavery, and has known only oppression all the days of his life."

What does coming home mean at different times of day
Does the open door greet ghosts or notes about supper
Have we caught a burglar in the act of flushing himself
out of mind, not having had time to rip the mattress
where our nap is waiting or was if the morning hadn't
told us point blank to make the future secure, which
we did in a variety of ways:

Mel the barber whose name should be Enrico
has been cutting hair for twenty
years or more and his prices aren't
barberlike anymore
but he still has his barber pole and piles of
US News & World Report, *Playboy*, *Field & Stream*, *Superman*
attracting hunters and moms.

The film has to be dropped at Fotomat with a coupon for
double prints; imagine, posterity having double the
opportunities to trace our genetic code; in this picture
and this one we're discussing something, drought or politics

Did someone need an abortion or did we want to know
a good restaurant; this man looks like he's out of luck
and the landscape prosaic, asking for scrutiny

Before we reach the post office we say no to two panhandlers
and yes to one—the lottery winner—and buying anything
seems criminal, felonies of the senses, of pure reason

Should everyone be what we were? Should they have
that perfect understanding? What about more hormones, less
brain? What about a mating season? If we can understand one
other, what then?

". . . Aristotle will defend abortion as preferable to infanticide. The Hippocratic code of medical ethics will not allow the physician to effect abortion, but the Greek midwife is an experienced hand in this field, and no law impedes her."

In this excavation, the one behind the oracle,
rusting in the kitchen midden, bay leaves also,
other herbs in a dusty urn, more rust
belonging somewhere else

No one, it seems, has written history beginning
I'm sorry, I'm sorry for, I'm sorry that
but someone better apologize for this mess
pretty soon, even though it's an accident, a hazard
life throws at us, a squall we're not ready for
but escape from most of the time

This is stuff for a paleontologist, history in rocks
what mother did when she was young, it takes awhile
to learn her imperfections, and now in her middle age
all of us kids with our well run lives
are out of control

"... the most brilliant civilization in history consumes itself in a prolonged national suicide."

The least I can do is bear witness
because I didn't help; instead, I ate
the remains of a fish, nicely done up
in a sauce of green mangoes with a fragrant
rice pilaf on the plate, too, and wine,
wine, of course

But I knew because we weren't allowed not to;
someone wanted our jobs, someone we'd
never met, someone was shooting, solving
problems, no doubt

The terminally young, in schools like
fish, never knew what they
were missing; meanwhile, back at the halibut
I enjoyed myself, imagined my lady's skirt
carelessly riding up her thigh, allowing the wine
to fool me into thinking of possibilities

That at one time, you know, a walk on the
boulevard under the stars, a butterfly against
a blue sky in summer, water from the lake spilled
on our lips

Well, I was there all right, recycling newspapers
with recycled news; what did I do, you may ask,

ready to throw me into the brown night that we see
every evening about this time

I saw it all on television, on faces of friends,
in litter in the park, I did nothing, but neither
did I know what to do

John down the street walked his dog, Bob called
about tennis, I was there and I swear this was
true, we let the sky fall, expecting someone,
no one we knew, to replace it

"The Greeks consider romantic love to be a form of 'possession' or madness, and would smile at anyone who should propose it as a fit guide in the choice of a marriage mate."

Take me take me take me
Grab my ass while they check our baggage
Shout I can't live without you when you're in crackers
and I'm lost in cleansers
Kiss me while the light turns green
and tell the cop my bald head is your magnet

In the Chinese restaurant I'll lift
your dress with chopsticks
In the coffee shop we'll drip grind
In the post office we'll lick stamps together

I'll fill the bathtub with pasta for you
We'll pesto, we'll alfredo, we'll marinara
We'll even carbonara

We'll defy telephones and vacuum cleaners
running toilets and rain coming in
overdue notices and yield signs

We may even ignore the kids

" 'Among the ancients,' said Stendhal, 'the beautiful is only the high relief of the useful.' "

The cowboy saw the flowers look like weeds
Not his territory, of course, the taboo colors
Hadn't said the word pink since he was five,
except in reference to pigs

Of course he could pick some, thin 'em out,
hand 'em to the little woman,
Shucks, he'd say, maybe

No question they were pretty purty,
which he had noticed himself when
he said, "Well, Amigo, think I'll
pick some a these for the little woman,"

And Amigo had said, "What fer? Is it
a holiday?"

"She'll be tickled pink," he said,
"Which is how I like her"

"Pink and tickled? Why not whiskey instead?
Then she'll be pickled pink. Haw Haw"

Daisies wuz the cowboy's favorite,
He could recognize 'em
she loves me, loves me not

He was a regular ol' romantic soul
a pinto pony, a sunset, a weeping
woman he'd left behind,

On her way on the Overland stage
the cowboy's wife to the cowboy's life
pink and blue bonnet and other
things all dream colored

Yella and red he could say,
black, brown and white came easy,
blue and green not too bad but not
simple like gray

He practiced saying orange
feeling his lips circling like a wagon
train in trouble

But pink stretched his mouth
like the barber looking for a bad tooth
and it made him wince for no good reason

Somehow saying this word was
like interrupting a woman in the privy,
blushed just thinking about it

Amigo said, "Looks like you
gotta bit of pink eye there; better
keep outta the light"

"I do?" said the cowboy, looking
 in the looking glass behind the bar
"I now pronounce you," someone said

 And the cowboy said, "Shucks."

"[The Athenian] has no patience with learned obfuscation,
and looks upon informed and intelligent conversation as
the highest sport of civilization."

Thought thickens my tongue and leaves me to mark where
the conversation has been while my wife and her daughters
race ahead, unafraid of blind alleys and interstate highways.
Because the daughters are young they say "you know" and "like"
to mark time and punctuate their explanations. Like they know a lot,
you know. I am serious about this.

I imagine that I've already told Paula about plans for the summer, my
reactions to news of the day, messages from the market place, so
when I tell her that I've made a decision she says about what and
we're both surprised. This isn't the way to keep a marriage fresh.

Through the last dozen years of my life I've listened to Presidents
speak in patterns only Chomsky can make sense of, and one was
even called the Great Communicator though his familiarity with life
as we knew it was coincidental at best. Before he was President,
he convinced many of us to buy Borax and products from General
Electric.
Thus we were aware of his ineffable intelligence.

My students fall asleep waiting for me to finish a sentence, the rest
no doubt refreshing them while I wonder why language has failed me
once again, my thoughts not pearls but the oysters themselves shut
tight. I'll never tell what I know even if I want to.

Sometimes I purposely call when they're not home, but I leave a
polite message, which I've already planned. The daughters call for

permission, hoping the mom is home but she's not. Can I help? Well, y'know like y'know Vicki was coming over? No, I didn't but that's ok. You didn't? Well I told Mom, anyway would it be okay if I like had

Bridget and Vanessa over instead? Vicki can't come. I see. Well, can I? We like wouldn't disturb you'n'Mom. How much homework do you have? Well y'know how me and Vanessa studied history together for the last test? You did? Yooo know, with the philosophy stuff, y'know John Locke and whatzizname y'know Rousseau, I got an A on it.

"These deities quarrel like relatives, fornicate like fleas, and share with mankind what seemed to Alexander the stigmata of mortality—the need for love and sleep; they do everything human but hunger and die."

Here in poems, where few think to look,
the answers, all we might ever need
language more powerful, richer anyway, than the people
it comes from, the ability to stop the sun
like Joshua, fix up the planet, put it
on the market

And of course there'd be buyers; didn't Tom Sawyer
find value in an apple core and dead rat; the
possibilities for a used planet are infinite,
everyone realizing they don't make 'em like they used to

Now the original architects, no one's sure what they
look like or whether they're still with us, spent years,
years making alterations: oceans, mountains, deserts,
the whole nine yards erected, renovated, moved

By the time they tired of the project, no end of plans
needed finishing
and they're off fishing I'm willing to bet,
washed their hands of the whole idea

So here we are with the for sale signs and a permanent
open house, no doubt hoping for rain, which, after which,
everything's washed and shiny

"The Greeks consider romantic love to be a form of "possession" or madness . . ."

I'd prefer we got on with it, the next
or even after that, the one daughter with
a broken heart, the other with
a blunt instrument. Most of the time
I can't tell which is which, both
of them so convincing with
their accusations.

In the gray and windy world of wet birds
in the sort of time when dry feet
make all the difference and afternoon
is an animal confused, unable to
come in out of the rain

desperate love, the only kind, really,
turns into junk mail trying
to convince us of righteous cause

Meanwhile, the sisters fight about a puppy
that's not even house broken; or maybe the
puppy is a swan or a bull or a flower. Now
that the girls have grown, instead of Spock
I pick up Edith Hamilton for advice

Amazingly enough, we're still trying
to figure out allowances despite
their unmade beds, their clothes strewn
midst plates of fossilized crumbs

"No absolute truth can be found, said Protagoras, but only
such truths as hold for given men under given conditions."

He may feel empowered to pick up an
automatic weapon—who's
to stop me, he thinks

From the train window you can see
    every junk yard in America,
all the corrugated iron rusting on the edge of town,
the chassis of a '65 Chevy Impala, stripped for parts,
rusting in weeds or snow, a piece of barbed wire winding
as ivy, and all of it means something—
    if you only knew what

The bells and lights of the crossing gate as
    you pass the mannequins
in the four a.m. dark. Why are they up and
I just want the truth, Harvey said to Joline
Every heart beat changes us; it dies a little

the train wants to sing the weather
the Mississippi's out of its banks
LaCrosse, Wisconsin looks for a hero
owls of thunder, a moth-eaten summer

Unfolding a lawn chair, sitting and waiting
shooting rats at the city dump with
his high school Spanish teacher
who has beer

They're dying of respect, and as for Joline,
well, I'd best keep that to myself—
she baked lies for supper, don't y'know
she shoulda flagged Amtrak at Winona

the bells and lights all my life, the train
without sinners strokin' east into
that righteous dawn come hell or high water
and high water comes like overdue taxes

she didn't even think of the train, she
took the bus

"The Athenians might be excused for giving him hemlock,
since there is no pest like a conscious logician."

this is the cup of mirth with which we try
this is the half glass of history that resembles gin
this is the demitasse of Ottoman philosophy

astronomy from Egypt has come via UPS
olive oil from Athens and figs from Judea
here are hollow bones from some long dead bird

these we learned by asking questions, which
we answered with more questions, which
we asked one time too many

imagine the excitement of finding a teacher
worth killing, ideas with the salt and smell
of sweat, the subversive eye all the way open

that implies knowing may and not knowing may not

here is a salad of mixed greens picked fresh
with brain and vinegar, oil and backbone,
bread and sinew, the teacher's words

here is the next day hungry again

Between Ice Ages *was designed and printed by Robin Heyeck. The frontispiece of the poet's front door was drawn by Barbara Hazard. The edition was sewn and bound by Cardoza-James.*